W9-BMM-861

Antique

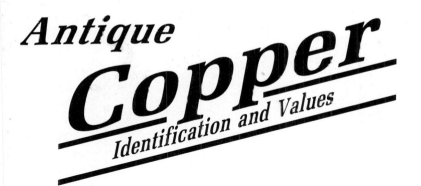

Copper
Identification and Values

Mary Frank Gaston

COLLECTOR BOOKS
A Division of Schroeder Publishing Co., Inc.

To Jerry and Jeremy

The current values in this book should be used only as a guide. They are not intended to set prices, which vary from one section of the country to another. Auction prices as well as dealer prices vary greatly and are affected by condition as well as demand. Neither the Author nor the Publisher assumes responsibility for any losses that might be incurred as a result of consulting this guide.

Additional copies of this book may be ordered from:

Collector Books
P.O. Box 3009
Paducah, Kentucky 42001

or

The Author: Mary Frank Gaston
P.O. Box 342
Bryan, Texas 77806

@9.95. Add $1.00 for postage and handling

Copyright: Mary Frank Gaston, 1985
ISBN: 0-89145-302-4

This book or any part thereof may not be reproduced without the written consent of the Author and Publisher.

Acknowledgements

I wish to thank a large number of people who contributed to this book in a variety of ways. First, I thank my publisher, Bill Schroeder, for his encouragement in this endeavor. Publishing the photographs in color truly makes the pieces come alive. You can almost see apple butter being stirred in a kettle over an open fire, or a warming pan being filled with hot coals to take the bite off of icy sheets!

Second, I thank my editor, Steve Quertermous, for the multitude of tasks he performs in transforming typed pages and stacks of photographs into a "real" book! He is always there behind the scenes.

Third, I thank my husband, Jerry, for his photography. Copper lends itself to natural settings, and Jerry spent much time fussing over pieces trying to place them in an appropriate background when possible. He also managed to stay out of the pictures pretty well which is no easy job since copper is like a mirror. Photography was not Jerry's only contribution. He edits the manuscript and helps in the total design. I am happy to say that he enjoyed this project as much as I did.

Fourth, our thirteen-year-old son, Jeremy, is quite adapt at proofing and finding typos! He is also teaching me how to use the computer and word processor which will be a big help on future projects.

Last, but certainly not least, I thank the many collectors and dealers who permitted me to photograph their copper and "talk copper" with them. Their contributions are truly the backbone of this book. Not only did they provide objects to be photographed, often ferreting out pieces which might have been obscured among a large group of items, but they provided informative, descriptive, and often colorful bits of information about particular pieces. To each of the following, I extend a very sincere "Thank You!"

Albert's Mall, Nashville, Indiana
Sallie Tucker Anderson, The Boll Weevil Antiques, Calvert, Texas
Anteaque Tyme, Dallas, Texas
The Antique Center, San Antonio, Texas
The Antique Connection, San Antonio, Texas
Antique Showcase, Fredericksburg, Texas
Antiques on Main, San Antonio, Texas
Antiques & Things, The Emporium, Bastrop, Texas
D. J. Blackburn, Waco, Texas
Joe and Von Bolin, The Tulip & The Bird, Fredricksburg, Texas
Joyce Brown, Prairie Wind Antiques, Fredrick, Oklahoma
Stella Brown, Migration Antiques, Bryan, Texas
Helen Buchanan, The Copper Lamp, Dallas, Texas
Peggy Burrows, The Emporium, Bastrop, Texas
Bob and Martha Davison, Horseshoe Antiques and Gifts, Fairplay, Colorado
Den of Steven Antique Mall, Louisville, Kentucky
The Depot Antiques, New Braunfels, Texas
The Drews, Fort Worth, Texas
Maizelle Dunlap, Bessie Mai's Antiques, Fort Worth, Texas

Eclectic Ideas, Dallas, Texas
Asa and Sue Ellis Antiques, Arlington, Texas
Donald J. Embree Antiques, Inc., Dallas, Texas
Linda Eulich and Isabelle Young, M.I.L.E. Galleries, Dallas, Texas
Puddin Evans, Mondays Antiques, Dallas, Texas
Rose Flocke, The Doubletree, Winberley, Texas
Virginia Fresne, Precious Memories, San Antonio, Texas
Dustin and Evelyn Gorden, The Barn Haus, San Antonio, Texas
David Harris Antiques, Big D Bazaar, Dallas, Texas
Erwin and Jamie Hendrix, Jamie's Antiques, Big D Bazaar, Dallas, Texas
Bill R. Lafferty, Bildor's Glasstiques, San Antonio, Texas
Maralee and Lauren Langholz, Mara Lang's Antiques, Mission, Texas
Elizabeth Leconey, Highland Park Antiques and Nauticles Ltd., Dallas, Texas
B. Lightsey, San Antonio, Texas
Roberta McCrary, Bois d'Arc, Calvert, Texas
The Main Place Antique Mall, San Antonio, Texas: Glenda Covington, Dorothy
    DeBona, Hazel Hamm, Ramona Huffman, Dick Klamm & Mark Linn, Bessie Long,
    Laura McLeod, Rose Mason, Helen Nelson, Tommye Newman, Maxine Ricter
Esther Maldonado, Esther's Antiques, Bandera, Texas
Martindale Antique Mall, Martindale, Texas
Ward and Don Mayborn, Uncommon Market, Dallas, Texas
LaDonna Mechaley, Serendipity Shop, Rapid City, South Dakota
Barbara & Thomas Morrison, The Victory Antiques, Dallas, Texas
Motif International, Dallas, Texas
Deanie Nolan, Deanie's Front Parlor, San Antonio, Texas
Rosie O'Reilly Antiques, Austin, Texas
Our Favorite Things, Dallas, Texas
Pendleton's Antiques and Collectibles, San Antonio, Texas
Judith Peters, Big D Bazaar, Dallas, Texas
Return Engagement, Dallas, Texas
Nadine Reynolds, Austin, Texas
Harvey Richman, Le Monde Antiques, Dallas, Texas
Faye Schoenfeld, Antiques and Interiors, San Antonio, Texas
Leola and Edgar Schulze; Edgar B. and Connie Schulze, Lee-Ed Antiques,
    Fredricksburg, Texas
Jim and Jerrie Shepard, Shep's Country Mouse, Belton, Missouri
Bud & Betty Sparks, The Attic, Bryan, Texas
Jean and Mike Sudderth, Country Cottage Antiques, Fredricksburg, Texas
Bobbie Terrell, Lost Pines Antiques, Bastrop, Texas
Al and Wanda Trigg, Al's Trifles & Treasures, Hurst, Texas
Mary and Ray Vaughn, Mary's Antiques and Collectibles, Louisville, Kentucky
Joe Webb and Jerry LaFevor, Paradise Antiques, Dallas, Texas

Preface

Old copper, attracting varied interests, occupies a vital niche in the collector's market. Copper's enduring and attractive qualities have caused it to play an important role in the world's history. Copper articles reflecting progress through time are treasured relics by an increasing number of people. Hand-made copper utensils and instruments made many hundreds of years ago are rare today; such examples are preserved mostly by museums. But an adequate supply of copper made during the last century and the early years of this century furnishes an intriguing assortment of items for today's collector.

This book focuses on those types of copper articles currently available in the antique marketplace: that is, items which can be found at antique shops, shows, antique malls, flea markets, or auctions. With few exceptions, items shown here were for sale by retail dealers and were not in personal collections.

A brief introduction to copper precedes the photographs. The inherent characteristics of the metal, relevant historical information, and collecting notes with an emphasis on prices and current reproductions are outlined. Terms, which are often used to identify certain copper objects but whose meaning may not be clear, are also included. The Price Guide at the end of the book quotes a value for each item shown.

About three hundred color photographs have been selected to illustrate the types of collectible copper seen most frequently for sale today. The photographs are arranged in three sections. The first section, household copper, comprises the largest part of the book. Examples are presented according to where such copper items might have been found in a home of the past such as the parlor and study, or kitchen and dining room, for example. This method of illustrating was chosen over an alphabetical approach because it seemed a bit more interesting. An Object Index is also provided in case items do not appear in the room you might expect them to be in!

The second section, commercial copper, includes trade related items pertinent to some businesses such as restaurants, newspapers, and even boot makers. Tools and instruments, used either in the field or in business, are also shown in this section.

The third section, current reproductions, has been emphasized in order to help collectors easily recognize several types of new copper found side-by-side with old copper at some antique outlets throughout the country. Some hints on differentiating the old and the new are also discussed in the first part of the book.

I hope you enjoy taking a step back in time to look at the nature and use of our ancestors' copper. Most of the articles are rarely used today for their intended purpose, but they furnish an interesting appreciation of the past.

Mary Frank Gaston
P.O. Box 342
Bryan, Texas 77806

Readers wishing to correspond, please include a self-addressed stamped envelope if a reply is requested.

5

Contents

Characteristics and Historical Background

Copper is a basic metallic element derived from mineral ores occurring naturally in the earth's surface. The metal is essentially soft, although it is stronger than gold. Copper is thus malleable and easy to shape. The metal is very attractive in its natural reddish-brown color, although its color may vary from pink, to a deep rust, or to a bright red. Its polished surface reflects light and takes on a pleasing glow and patina through use and polishing over time.

Copper does not rust, but it can corrode, especially on contact with acidic substances. Verdigris, a blue-green coating, develops on copper when it is exposed to the elements over long periods. That type of discoloration is evident on buildings with copper roofs. The verdigris can be removed, but it will eventually return. The coating, however, does not harm the metal seriously.

Cleaning copper does not require an inordinant amount of effort, nor must it be done frequently. Commercial copper polishes quickly restore a warm sheen that will remain for some time. Also, pieces can be protected by a coat of lacquer. Then they need only periodic dusting. Many collectors of old copper, however, usually prefer the natural patina which pieces acquire over time.

Copper is very durable. It really does not wear out. When objects do become worn, they can be patched or reseamed. Alternatively, the worn pieces can be melted, and some new item fashioned from the scrap metal.

Copper is a good conductor of heat. Centuries ago, copper vessels were used for cooking purposes. Copper cooking ware remains popular even today. In modern times, the metal was discovered also to be a good conductor of electricity, being used extensively in electrical wiring.

Copper is an essential ingredient in manufacturing several metal alloys. Those alloys are stronger than copper, and thus greatly diversify the use of the metal. Bronze, made by combining copper with tin, is thought to be the first man-made alloy. Copper and tin, in different proportions than those used to make bronze, produce other alloys such as bell metal, gun metal, pewter, and Britannia metal. Brass, composed of copper and zinc, appears to be the most widely used copper alloy. Because of copper's characteristics, it is easy to see why the metal has played such an important role in the development of modern civilization up to present times.

The metal has been known from very early times in all parts of the world. Copper and gold are, in fact, considered to be the first metals known to man, although history cannot pinpoint the date or the loca-

7

tion of the first discovery of either. Archeological sites have yielded evidence that copper was used by the ancient civilizations of China, Egypt, Greece, and Rome. Copper tools, weapons and cooking vessels were common in Europe during the Middle Ages. North American Indians possessed copper-made articles when the continent was discovered by Europeans.

Although copper's history can be traced to prehistoric times and to various parts of the world, relevant historical information for today's American collector centers for the most part on English and American copper. England was the largest source of copper-made articles for American colonists. England's copper industry, however, did not develop on any large scale until the beginning of the 18th century. Previously, England had relied on imported copper from Germany and Holland, either in the form of ready-made items, or in the raw material which could be worked into objects.

In 1689, the Mines Royal Act took control of the mines away from the Crown, and as a result, it became possible for individuals to own mines and engage in metal manufacturing. Large deposits of copper were available in England in the region around Cornwall. From the early 1700s through the first part of the 19th century, Cornwall was the center of English copper production.

Progress in various types of industries took place in England during the 18th century. Improvements in mining ores and many innovations in manufacturing metals occurred which helped to ease and quicken the process of making items from copper and its alloys.

Copper was not an easy metal to extract from its ore. The ore itself was difficult to mine, and once it was brought to the earth's surface, it had to go through several processes before it was properly refined for use. The ore had to be "dressed," which consisted of separating the usable ore from the unusable. Many tons of ore could be mined which actually might produce very little copper. After dressing, the ore was smelted. During the smelting process, arsenic was burned off as a by-product. The fumes were poisonous and hazardous to both workers and the surrounding area. For that reason, smelting plants were located some distance from towns and villages.

Smelted copper was poured into slabs or ingots which could be refired, melted, and worked by hand. The copper was hammered into variable sizes and thickness to make different objects. Some articles were completely beaten or hammered into shape while others were made from separate pieces joined together by soldering. Copper was rarely cast, even though casting was widely used as a method of shaping objects made from copper alloys such as bronze or brass. Copper did not have enough strength to be cast into most items.

During the 18th century, it became possible to roll metal into sheets by machines. The sheets could be cut into precise sizes resulting in more uniformity for shaping articles. Dovetailing was a method used to join sheets of copper. A piece of copper was cut with indentations, and another piece was cut with projections which would fit together. A soldering metal joined the two parts. The soldering metal had to have a lower melting point than copper. Lead, tin, and brass were used. The soldering was visible because of its different color. Pieces might be dovetailed on the side, around the body, or on the base.

Dovetailed joints were originally cut by hand. They were usually large and widely spaced during the 1700s, and the cuts were often irregular. During the 19th century, dovetailing became more exact. The metal was notched by machine and a regular pattern was produced. Simple straight seams began to replace dovetail construction during the 1800s.

Another method of shaping metals was introduced during the early 1800s. This was a spinning process where objects were made by using a die and rotating device which produced objects without any seams. The pieces were also much lighter in weight.

In England, in 1769, a procedure was patented for stamping designs on metals. Later, stamp and die techniques were adapted to shaping whole small objects such as hardware. By the beginning of the 19th century, larger items could also be shaped entirely by stamping. England's metal industry was, in fact, almost totally mechanized by about 1800.

Other important inventions in connection with copper took place in England during the 18th century. Especially notable was a new method of tinning copper. Some foods, in direct contact with copper, created a chemical reaction that resulted in a tainted taste and even food poisoning. Lining pans with tin solved the problem. Tinning prevented the tainted taste, but unfortunately, lead had to be used in that process. Lead was necessary as a flux to make the tin adhere to the copper. Lead, of course, caused a slow form of poisoning. A method of tinning without using lead was sought for many years; it was finally discovered about 1756 (Wills, 1968). During the 20th century, stainless steel largely has replaced tin for lining copper cooking ware.

Plating copper with another metal was possible after 1742. Thomas Boulsover, working in Sheffield, England, introduced the technique. A sheet of copper was placed between two sheets of silver and heated until the metals fused together. The plated copper was then rolled into thin sheets which could be fashioned into different objects. Sheffield plate became the name coined for items made in that manner. Objects had the look of solid silver, but the cost of manufacturing, and hence the cost to the customer, was considerably less. Consequently, another important contribution was added to the long list of uses for copper.

Electroplating replaced Sheffield plating toward the middle of the 19th century. That technique of plating one metal with another by means of a chemical process called electrodeposition was discovered by the Elkingtons, another English company. Electroplating was easier and less expensive than Sheffield plating. Not only silver, but gold and other metals have continued to be used to electroplate copper. Today, many items which were once electroplated are stripped to the copper base metal. Silver plated pieces show wear over time and are unattractive. Replating is expensive and is not permanent. By stripping away the silver or other metal, a perfectly fine and lasting piece of copper will be uncovered! Businesses throughout the United States specialize in stripping metals today.

There was little American copper manufacturing until after the beginning of the 19th century. By English law, American colonies were forbidden to engage in any form of manufacturing in order to provide a ready market for the mother country's goods and also to serve as a source of raw materials. Such laws, of course, were difficult, if not impossible to enforce. Records show that various individuals did work as coppersmiths before independence was won in this country. For the most part, however, the coppersmiths worked with scrap copper rather than imported copper ingots or native copper.

Paul Revere, America's most famous coppersmith, was working with copper in the last quarter of the 18th century. He imported copper from England and later developed methods to improve the use of scrap metals. Examples of his work are museum pieces today. After 1800, however, his business focused on commercial and industrial uses of copper for construction.

During the early 1800s, other foreign sources of copper, in addition to those in England, became available which enabled American copper manufacturing to really grow as an industry. Although copper was available in the colonies, the metal was not mined to any great extent until around the middle 1800s when vast deposits became available from the western mountain regions. It had always been cheaper to import the metal, and even with large accessible deposits, that has continued to be true even today.

Availability, Pricing, and Reproductions

There is a large demand for "old" copper of all types. "Old," however, is a relative term. When used in relation to copper available today, the word does not usually imply "ancient," or even several hundred years. Old copper most often refers to pieces made after 1850 until about 1930. Examples from that period are in greater supply. Many items made during that era are also truly relics of the past in the eyes of young collectors, and they are nostalgic memories for many others.

The large shipments of antiques imported from England and other countries for some years now often contain copper. Occasionally, some pieces from the 18th century or early 19th century are included, but most pieces date from a much later time. When such early copper does surface, the price can be quite out of reach for the majority of collectors. Copper, in general, is much more scarce than other antiques such as furniture, glass ceramics, or even other metals like brass or silver. Just take a look when browsing at shops, shows, or flea markets, and that fact will be quite clear. Copper enthusiasts must diligently search to find interesting pieces.

"Old copper" brings to mind basically utilitarian articles, simply or even crudely made, with little, if any, decoration. A lot of copper does fit such a description. Highly decorative copper, such as jardinieres, vases, and wall plaques, can be found, however. Many purely functional copper items were also decorated as well.

Pieced or punched patterns and randomly hammered work are perhaps the simplest forms of decoration found on copper. Engraved designs, made by cutting into the metal with a sharp tool, and embossing (repousse), achieved by hammering designs into the metal from the reverse side, can be quite elaborate. The ability to stamp patterns on sheet metal greatly widened decorating possibilities for machine made pieces, allowing very detailed and intricate decoration. Gilding and enamelling also have often been used to decorate copper. Other metals such as silver, pewter, and brass were combined with copper simply because the combination of the two metals was attractive. Some examples of those different decorating techniques will be seen in the photographs.

Decorative copper can reflect definite shapes and designs in vogue during certain historical eras. Pieces may exhibit Georgian lines or Art Nouveau decor, for example. Those periods which are most often used to identify particular styles are:

> **Arts and Crafts**--a movement during the late 19th century, circa 1875, emphasizing original designs and hand-crafted construction. The trend was initiated in response to the large growth of machine-

made, mass produced styles predominant during the middle Victorian years. (For an example, see plate 24)

Art Deco--a style focusing on geometric lines and stylized designs, popular from circa 1925 until 1940. The name comes from the 1925 Paris exhibition, *Exposition des Arts Decoratifs et Industriels Modernes*. (See plate 22)

Art Nouveau--literally a new form of art, in reaction to elaborate, over-worked Victorian designs. Art Nouveau styles were characterized by smooth flowing lines and naturalistic subjects. The new trend was popular for only a short time, circa 1890 until 1905. (See plate 21)

Georgian--several styles, marked by a stately, classical line, introduced over a long period, circa 1714 to 1830, when England's kings were all named George (George I through George IV). (See plate 18)

Victorian--very ornate, rococo styles popular during the reign of Queen Victoria, 1837 to 1901. (See plate 6)

Among collectors, American-made copper items are highly prized over foreign-made copper. It is difficult to attribute copper articles to specific origins, however. Not only did the early makers in Europe or America seldom mark their wares, but types of objects made and styles of pieces were also quite similar. English manufacturers copied German or other European designs, and Americans, in turn, copied English examples. Moreover, the types of items and the styles did not change very much over time. As a result, dating is rather inexact.

A mark, or a name, on a piece of copper does not necessarily indicate a particular coppersmith. The name or initials may only refer to the owner of the article. It was not uncommon for individuals to have their name engraved on metal wares. The objects were treasured possessions. They were cared for, mended when necessary, and handed down to succeeding generations.

Unless the name on a copper item can be matched to a recorded coppersmith, the manufacturer cannot usually be proved. If a mark includes a location, the country of origin is not difficult to determine. Marks which do not include locations, however, cause the origin to be uncertain, unless the name can be matched to a known coppersmith of a particular country. Relatively few such names have been recorded, and even they may not have marked all or any of their wares.

One American writer, Henry J. Kauffman, has compiled a list of American coppersmiths and the years during which they worked. He shows approximately 140 names of persons who were known as coppersmiths working between 1740 and 1863 in Connecticut, Maryland, Massachusetts, New York, and Pennsylvania. (See his "Early American Brass and Copper and its makers," pp. 104-107 in Albert Revi (ed.), *Collectible Iron, Tin, Copper & Brass*, 1974). On today's market, however, few pieces surface with one of those signatures. If an old piece does have a mark, it would be worth the time to see if the name does appear on Kauffman's list. It is not impossible for rarities to crop up now and then.

Manufacturer's marks were used more frequently, both here and abroad, after 1900. Thus marked examples from after the turn of the century are more common. They are not considered "rare" in the same sense as 18th and early 19th century markings. Few pieces illustrated in this book were marked. For the marked pieces, the prices generally reflect their degree of rareness.

Most of the copper featured in this book dates from the last quarter of the 19th century through the first quarter of the 20th century. Dates have not been included in the captions of those photographs. The term "19th century," however, has been used to indicate a period earlier than the late 1800s; similarly, "20th century" has been used to indicate a time later than 1925.

The majority of copper in this survey was also of European or American origin, as opposed to Oriental, Far Eastern, etc. Country of origin has been included in the descriptions of the items where it was reasonably possible to determine. For instance, recent European imports are obviously European rather than American. Some of those are also identifiable as English rather than French, etc., or vice versa. After 1890, United States tariff laws required foreign countries to mark wares exported to America. Therefore, distinguishing American-made copper from European-made copper became less problematic. The inclusion of a country's mark on a piece of copper also signifies that the item was made after 1890.

Current prices for antique copper have been influenced by several factors. The base price of copper increased sharply during the early 1980s. The high price of raw copper encouraged some people to round up their bits and pieces to be sold for scrap. Others began to stash away the metal in all of its forms, including pennies. Some banks even found it necessary to pay a premium if people would cash in their pennies because of the scarcity.

Higher copper prices and the resulting hoarding caused prices to escalate for all types of manufactured copper. That increase was also reflected in the price demanded and paid for old copper. For example, copper wash boilers which sold for $40.00 to $50.00 in the late 1970s can seldom be purchased for less than $80.00 to $100.00 today. That same type of increase can be seen for numerous other articles of collectible copper.

A widening interest in kitchen collectibles, primitives, and tools has caused prices to remain high for copper. Interior decorators and home decorating publications have helped to acquaint the general public with the warmth and charm old copper furnishes as accents and accessories for any type of home, ultra-modern to log cabin style. Warming pans hang by the fireplace, a variety of molds line a kitchen wall, and coal scuttles and preserving pans hold magazines or kindling wood. Today, interest in old copper is by no means confined to antique and metal collectors.

Raw copper prices have stabilized, but prices for collectible copper have not decreased. Bargains are hard to find. Late 20th century American all-copper pennies may, in fact, be the most affordable of all copper collectibles. Since 1982 pennies have been made of zinc and only coated with a thin layer of copper. It began to cost nearly a penny to make a penny! American zinc pennies may even become extinct. England discontinued minting her coppers several years ago, and the same has been suggested for our coin system.

Prices are not likely to decrease for copper in the future. Even mid-twentieth century pieces are attracting the collector's eye. Now, they are a good bit less in cost than turn of the century items. Many, however, continue to treasure and search for much older examples, finding cost secondary to owning an authentic part of the past.

For the majority of copper articles illustrated in this book, values quoted in the Price Guide reflect the seller's tagged price for the item. In those instances where prices were not available, or if pieces were in personal collections, an average price was derived from retail prices of like objects. Although some pieces were priced either considerably higher or lower than a similar item, on the whole, there was not an extremely wide differentiation in prices among categories of objects. Thus, it is possible to get an idea of the range of price certain examples of copper command today, such as coal containers, kettles, wash boilers, etc. Unlike ordering new merchandise from a catalog for set sums, antique prices are influenced by a number of factors including scarcity, size, condition, collector interest, dealer knowledge, regional markets, and type of selling outlet.

An unwelcome by-product of mounting interest and high prices for old copper is the rampant deluge of reproductions of a number of antique and collectible copper items. Through the ages, copper, of course, has been copied, as have other metals such as silver, bronze, pewter, and brass. Copies of 17th century pieces were made in the 18th century, and 18th century copper was copied during the 19th century. Those copies have now been around for quite a number of years. Thus, they have become legitimately "old," meriting their own antique value and interest.

N.H. Moore gives an enlightening account of copper reproductions made in New York City.

In Allen Street you will hear the sound of the metal-worker as he swings his mallet, and if you are allowed to penetrate the dusky recesses of the back shop you will find at work a swar-

14

thy man with dark eyes, and hanging around him are shears and pinchers, hammers and mallets, sheets of copper and patterns by which to cut out his metal. He works at a long rough table, and near at hand is a crude furnace at which he heats his metal, and when it is at the proper temperature to make it malleable, he begins to hammer it into shape, stroke by stroke. As it slowly takes form you see the graceful shapes you admire growing before your eyes, with the hammer-marks which are always so esteemed as showing the work to be hand-made rather than machine-made. To suit 'the trade,' some of these newly made goods are battered and dented, and hung in the smoke to darken (p.167).

Do not start looking for this shop. The author was writing around the turn of this century! *(N.H. Moore, Old Pewter, Brass, Copper & Sheffield Plate* (1905), 1933) No doubt many of those copper items now make up part of today's collectible market. Those new pieces were the concern of Moore in the early 1900s, but today the objects are genuinely old.

Like Moore, today's collector must be concerned with currently made reproductions. Unlike those "new" items made in the early 1900s and described by Moore, today's repros are not hand-made. They may be advertised as the "antiques of tomorrow," but I am quite skeptical about whether they will be able to hold up to the test of time as well as those earlier hand-made copper reproductions.

Current copper "repros" should not be confused with other new copper which may be made along traditional or antique lines such as kitchen wares, lighting fixtures, and fireplace equipment. That form of new copper is usually well made, often marked, and sold at various retail outlets such as department stores, hardware shops, or mail order outlets, places where new, rather than old, merchandise is expected to be found.

The repros, on the other hand, are sold almost exclusively to the "antique trade" or some other retail business such as gift shops and restaurants specializing in a country store or home-style cooking atmosphere. The repros have been made for several years. They are manufactured in various foreign countries such as Taiwan, Japan, and Korea, and imported to this country by a number of entrepreneurs. The importers advertise the products as reproductions. They sell the pieces at very low wholesale prices in general. Pieces are even marked with the country of origin, but the mark is only a stick-on label, easily removable.

Although the businesses are not misrepresenting their merchandise or pricing items as though they were old, they do restrict their customers to "the trade," that is retailers who, by and large, are antique and collectibles dealers. To purchase from their catalogs or

15

showrooms, dealer identification must be presented. Consequently, where are those new pieces of copper going to be sold? Of course, they are found mostly where one expects to find authentic antiques and collectibles.

From the tremendous business the importing firms conduct, it is apparent that there is a large market for reproductions of all kinds, including copper, as shown by the number and variety of items which are available from those sources. Genuine antiques are getting harder and harder to find as well as more expensive each year. More people are aware of the value of their possessions, and rarely make the mistake of parting with a treasure for a small sum. I surmise that is why many dealers purchase the repros. They help fill in their stock and are easy and inexpensive to buy. Persons wanting a few pieces of copper for decorating a home may go to an antique shop to find it. Because copper is scarce, however, they may not find what they want, or the price may be more than they want to spend, thus they purchase a new repro to achieve the "effect." Thus, because the repros do sell, dealers continue to stock them.

Many dealers handling repros will tell the customer that the copper is new, but some may not. It is not uncommon for new copper to be purchased as old. Prices have a way of edging up each year so that the gap between prices for old copper and new copper narrows some year by year. The price tag is still usually a good indication of whether the article is old or new. With even a hundred percent mark-up, most repros are well below the cost of the same piece made fifty or more years ago. A copper bedwarmer for $25.00, a jelly kettle for $35.00, or a ship's lantern for $27.00 immediately identifies the piece as new, not a "steal."

When in doubt about whether copper is old or new, inspect objects carefully. Look for the patina on unlacquered pieces. If the copper has been lacquered, ask the dealer when the finish was applied. Many dealers have all of their old copper lacquered because some people prefer it. If the dealer can only tell you that the piece was lacquered when purchased, then it is indeed possible that the item is new. New coal hods, umbrella stands, bedwarmers and tea kettles have a lacquered finish. Some of the cooking ware does not, however. The reproduced kitchen wares, however, are not tin lined.

Repros are usually much lighter in weight than old copper. Some new pieces, such as lamps, may be weighted with lead. Copper is rolled very thinly for constructing the cheap imports. Edges are sharp, and pieces may not be finished off very well. New copper dents easily. "Distressing" is not uncommon, but it is usually so overdone that it is easily recognized as a ploy to make an item look authentically worn.

Dovetailed construction is really too expensive to be copied. Simple seams should not show an overabundance of the soldering metal. Copper repros may have a tin colored soldering abundantly dripping out of the seams. Another trick to "age" copper is an applied blue-green color randomly dappled on items such as weathervanes. The color is supposed to resemble the verdigris copper acquires through exposure to the weather. Some new pieces, which would never have been exposed to the elements originally, also have this coating!

Handles and feet on copper reproductions strive to copy those used years ago. Old iron handles may be attached to new pieces, or new iron handles with a fake "rusting" can be added. Brass paw feet, and handles, fashioned with lion heads and a ring through the nose, are commonly seen on a number of repros. The ceramic "delft-style" handle on a variety of copper coal buckets is a mark that the piece is not very old. Some of those new buckets have been around for several years now, but they are still being made.

Beware of marks found on some copper items. Identifying names are added so that articles may look as though they were made for certain businesses such as a hotel or railroad, or that they were made by a specific firm. The names are merely decorative. Markings embossed in large letters should not be difficult to determine as new. A copper and brass spittoon with "Wells Fargo" on the front is just one example. Other names stamped on small brass plates and attached to items such as bedwarmers and lanterns can cause more confusion.

Current copper reproductions may not even be made entirely of copper. Three different dealers showed me the same candy mold as an example of "old" copper. The piece is new and only copper plated iron. Be sure and take along a magnet when searching for copper. It can be very handy!

A few of the most common copper repros are shown at the end of the book. Hopefully they will help warn collectors about the types of new items which may crop up at various antique locations today. In addition, they should serve as a record which will alert future collectors to the copper reproductions made during the 1980s. In a few years, such items will have begun to acquire their own "patina" from handling, dust, and dents, as well as prices closer to those asked for the real thing!

Distinguishing and Naming Copper Objects

Several different names are often encountered for the same copper objects; and, the same name is frequently used for two or more objects. Some similar looking items often have very different names. This is true especially for kitchen related wares. A pan is not just a pan, for example. It may be a cauldron, a preserving pan, or a sauce pan. Names for articles have evolved through time based on how the piece was generally used. The same type of objects could have been used for different purposes in different countries or by different groups of people in the same country. Thus, it is not uncommon for more than one word to be found for similar items. Pails and buckets are basically the same object so are skillets and frying pans.

Sometimes names have strayed away from the intended purpose of the piece. A tea kettle is actually any type of kettle used to boil water or any other liquid. Tea is not made in the kettle, and the kettle's purpose is not limited to boiling water for tea. "Kettle" originally meant any large, deep pan used for cooking over a fire. When the custom of drinking tea was introduced, "tea kettle," an entirely different type of object, identified its purpose. Kettle, however, is still used to identify either one.

Some of those confusing names for copper objects are described here. Other words whose meaning might not be clear and which have been used to describe some of the copper illustrated are also included. The number in parentheses at the end of the description refers to the photograph number of one example of that item.

Basin--a bowl, more shallow than a tub, for washing hands, laundry, etc. (210)

Bed Warmer--a metal container, round or oval shaped, with a small neck and lid. The containers were filled with hot water and placed in the bed. They are later than Warming Pans. (205)

Chamberstick--a candle holder with a large flat base and attached handle. They were designed for lighting the way to bed. (199)

Cistern--a water container, usually covered, equipped with a spigot. (211)

Cauldron--a large kettle.

Can--a tall, cylinder-shaped container with a narrow neck. (120)

Coal Box--basically any covered coal container. (15 & 17)

Coal Carrier--a covered pan with a long handle. They were made to carry hot coals from one fireplace or stove to another. The top may have punched holes. They can be confused with Warming Pans. (54)

Coal Hod--an open coal container, found in various shapes and sizes. They may also be called a bucket or a pail. (Not illustrated)

Coal Scuttle--a coal container usually having part of the top covered. They may have a ring attached at the back for holding a small shovel or scoop. (13)

Colander--a round pan, pierced over all or most of the bottom, for draining liquids from foods. (59)

Fender--a horizontal piece of metal, only several inches in height, placed along the length of a fireplace to hold back the ashes. (12)

Geyser--a gas-powered hot water heater. (215)

18

Jardiniere--the French term for a planter and stand. The pot alone, however, is often referred to as a jardiniere. (7)

Jug--a fat, round pitcher (or a cylinder-shaped container with a small neck and handle). (123)

Kettle--a utensil for boiling foods or liquids. Kettles are deep, and they are larger than Sauce Pans. Handles may be attached to the side, or across the top (bail-type). Except for Tea Kettles, most are shaped similarly, but sizes vary greatly. Early ones were used for cooking over an open fire, and later ones were used on top of the stove, rather than in an oven. A few varieties are listed here:

Apple Butter Kettle--a very large kettle for making preserves, usually unlined. (68)

Fish Kettle--an oval-shaped pan for cooking fish, usually has an insert and a lid. (132)

Jelly Kettle--similar to a Preserving Kettle, but shaped like a pail. (76)

Preserving Kettle--any kettle for making preserves, often called a Preserving Pan. (70 & 72)

Stewing Kettle--similar to a Preserving Kettle, usually tin lined. (78)

Tea Kettle--a utensil with a spout, handle, and a lid, for boiling water. (155)

Lantern--an enclosed lighting fixture which may be suspended or carried. (4 & 266)

Lavabo--a basin for washing hands, usually placed below a wall-mounted cistern. (210)

Log Container--a large open container for holding logs. They may be highly decorated. Some look similar to large planters, and they are often sold today as such. (18)

Loving Cup--an oversized cup, originally intended for ceremonial purposes where the cup was passed around, and each person sipped (wine) from it. Through time, its use has evolved to a form of trophy. They may have several handles. (37)

Measures--usually cup or pitcher-shaped, often marked to indicate specific amounts. Hot Liquid Measures have a long handle. (87 & 88)

Molds--used to shape food into various forms and sizes, primarily used for eye-appeal. They range from simple to quite elaborate in design. Similar looking molds are referred to as Jelly Molds, Pudding Molds, Vegetable Molds, or Cake Molds. They can be used for any of those types of foods. Jelly Mold however, refers to any mold used for a gelatin-based mixture; it does not necessarily mean a breakfast jelly. Candy or Chocolate Molds are made in the form of a pan with small individual shapes for forming pieces of candy. Cake Molds often have a center tube, and they are deeper than a cake pan. (94 & 101)

Pail--a cylinder-shaped container with a bail handle, no lid, similar to a bucket. (85)

Pans--a broad term applied to a number of different types of cooking utensils.

Baking Pan--any rather shallow pan used in the oven. Cake and bread pans usually do not have handles (154); roasting pans do. (149)

Candy-Pan--a large round pan with side handles. They are not as deep as Preserving Kettles, but they are sometimes referred to as Preserving Pans or Kettles. (69)

Ebleskiver Pan--a pan with several cup shapes in the bottom. These are for making Danish dessert dumplings. Ebleskiver Pans are often confused with Escargot Pans. (145)

Escargot Pans--similar to an Ebleskiver Pan, except the cups are smaller. (147)

Frying Pan--a round, rather shallow pan with a long, straight handle, designed for cooking over a flame, or on top of a stove. (144)

Preserving Pan--the same as a Preserving Kettle.

Roasting Pan--a deep pan for cooking meats in the oven. (149)

Sauce Pan--a deep, round pan with a long handle and a lid. Sizes vary, and the pans were used for cooking any food that needed to boil or simmer. (134)

Stewing Pan--the same as Stewing Kettle.

Peat Bucket--an open container for holding blocks of peat or turf used in building a fire, similar to a coal bucket. (19)

Pitcher--a cylinder-shaped container for liquids, having a spout and a cup-style handle. Sizes vary from small cream pitchers to large ale tankards. (121)

Planter--any deep receptacle for holding a plant. Originally, the containers were designed for actual planting, but today the term commonly refers to a decorative holder for clay-potted plants. Planter and Jardiniere are used interchangeably. (26)

Plaque--a decorative dish designed for hanging. They are often called Chargers. (30)

Plate Warmer--a shallow pan fitted with a plate or platter on top. The pan has a small opening which can be filled with hot water to keep food warm. (182)

Pots--a broad term, like Pans, applied to a number of cooking vessels. "Pot" or "Pan" often refers to the same object. Pots are deeper than pans.

Coffee Pot--pitcher-shaped with either a short, open spout, or a long narrow spout. Sometimes, the short spouts may have a hinged lid cover (43). Perculator Coffee Pots often have a glass knob on top. (49)

Stock Pot--a large, deep, cylinder-shaped container, with or without a lid, side handles. They may have a spigot. (125 & 126)

Tea Pot--usually shorter than a coffee pot, with a long spout, handle, and lid, used for brewing tea.

Samovar--a large covered container, equipped with a spigot and some form of heating to keep the beverage warm; similar to an Urn. (179)

Sconce--a wall-mounted light fixture. (2)

Skimmer--a utensil for separating one substance from another; sizes vary; some are finely pierced all over, like a sieve or filter for separating milk; others may be in the form of a flat bowled ladle for separating solid substances, like fat, from a broth or stew; another type has a pierced bowl. (63,67,66)

Skillet--a Frying Pan. (Originally, the skillet was a three-legged pot with a long handle, made for cooking over an open fire.) (144)

Steamer--a fairly deep pot with a lid. They may have a tray to hold the food above the water. For steaming, only a small amount of water is used, in order to cook the food quickly to a just-tender consistency. (129)

Tea Kettle--see Kettles

Tinder Box--a round cup-shaped container used for starting a fire. Pieces of charred cloth, flint, and steel were kept in the box. A lid, made with a candle holder, fitted on top. Once the fire was ignited, the candle could be lighted and placed in the holder (60). Tinder Box may also refer to larger containers for holding pieces of kindling material. (12)

Tub--any large, deep container used for bathing or laundry. (221)

Urn--a large, covered container for serving hot liquids, equipped with a spout and some means of heating. Early ones had spirit lamps; later ones use canned heat, or electricity. (Urn also refers to a covered vase for holding the ashes of the cremated.) (179)

Warming Pan--a covered pan with a long iron or wooden handle. The pan was filled with hot coals to warm the bed before retiring. Some pans have pierced tops. They are commonly called Bed Warmers. (202)

Wash Boiler--an oval-shaped tub for washing clothes or boiling water. It may have a lid. (224)

Household Copper--Entrance & Foyer

PLATE 1. Outdoor Lighting Fixture, Porch Lantern, 22″h, cage-style construction, Victorian.

PLATE 2. Outdoor Lighting Fixture, Sconce, 12″l, brass trim, hammered body.

PLATE 3. Outdoor Lighting Fixture, Sconce, 9″l, pressed glass globe, lacquered.

22

PLATE 4. Outdoor Lighting Fixture, Hanging Lantern, 9½"h, celluloid shade, pierced work on top, probably Oriental.

PLATE 5. Outdoor Lighting Fixture, Gas Porch Lantern, 24"h, English.

23

PLATE 6. Mirror, 31"l, 25¼"w, Copper Frame, 6"d, ornate embossed designs, Victorian.

PLATE 7. Jardiniere, 28"h overall, Planter is 25¼"d, repoussé decor, Victorian.

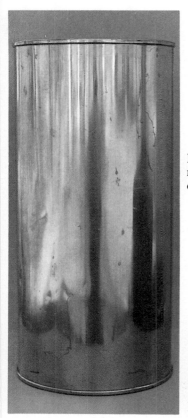

PLATE 8. Umbrella Stand, 19½″h, lacquered, made from an early 20th century American fire extinguisher.

PLATE 9. Jardiniere, applied brass decoration on planter, ornately shaped three-legged brass stand, English, Victorian.

25

PLATE 10. Planter, 15¼″h, 13¼″d, brass lion heads decor, brass paw feet, English, 19th century.

Household Copper--Parlor & Study

PLATE 11. Fire Screen, 21"h, 22"w; Fireplace Tools for coal fires, English.

PLATE 12. Fire Screen, 29"h, 27"w; Fender, 53½"l; and Tinder Boxes made into seats, scenic decor in high relief featuring a dog and her pups, Scottish.

PLATE 13. Coal Scuttle, 16"h, pedestal base, tin lined, lacquered.

PLATE 14. Coal Scuttle, 18"h, hammered body, embossed floral designs on top, Victorian.

PLATE 15. Coal Box, 18"h, brass medallions mounted on sides, brass paw feet.

PLATE 16. Coal Box, 17"h, (liner not shown), shell and ring handles, marked "British Made."

PLATE 17. Coal Box on wrought iron stand. Box is 22″h, 26″w, 11½″d, lacquered, designed in Art Nouveau style, English.

PLATE 18. Log Container, 27″h, 22″w, hammered surface, brass paw feet, Georgian, early 1800s, English.

PLATE 19. Coal or Peat Bucket, 10"h, 9"d, dovetailed construction, possibly Dutch.

PLATE 20. Coal Box, 16"h, 16"l, footed, urn finial, brass fittings, English.

PLATE 21. Stove Insert, 26"h, 20"w. repoussé work in Art Nouveau style.

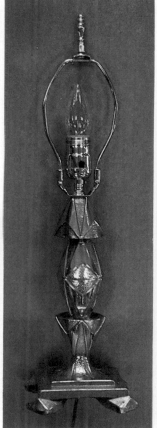

PLATE 22. Table Lamp, electric, marked "1928, Armour Bronze Core," lacquered, Art Deco lines.

PLATE 23. Plaque, 14″d, fluted border, embossed roses.

PLATE 24. Coffee Table, faux bamboo handles and legs, Art and Crafts period.

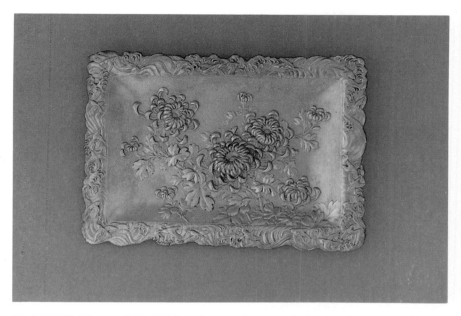

PLATE 25. Plaque, 12"l, 8½"w, chrysanthemums in high relief, mid-1900s example of copper craft art.

PLATE 26. Planter, 4"h, 5½"d, marked "China."

PLATE 27. Planter, 9″h, 10″d, brass handles and paw feet.

PLATE 28. Pair of copper and brass vases, 12″h, red jewels, Grecian design, mounted in allegorical figural bases, Victorian.

35

PLATE 29. Pair of Vases, 13″h, embossed Egyptian motif, Art Deco period.

PLATE 30. Plaque, 26″l, 22″w, repoussé figural and floral decor of Art Nouveau era, English.

36

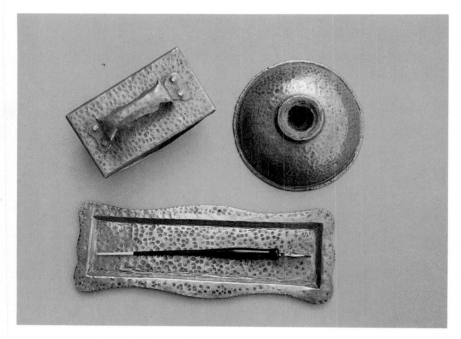

PLATE 31. Desk Set: Blotter Holder, Inkwell, Pen Tray, hammered surfaces.

PLATE 32. Wall Pocket Letter Holder, 9″h.

37

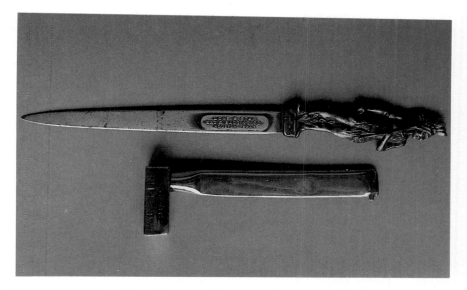

PLATE 33. Advertising Souvenirs: Hammer, 8″l, marked "Hale Fire Pumps," and Letter Opener, 12″l, marked "Lone Star Bag and Bagging, Co., Houston, Texas."

PLATE 34. Inkwell, hammered copper and cast brass, Art Deco design.

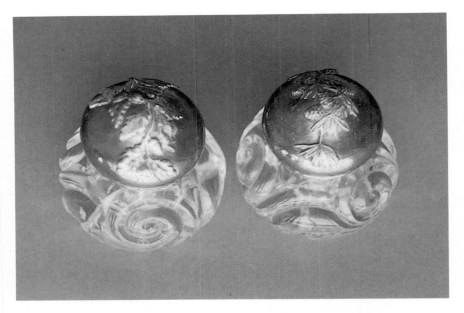

PLATE 35. Pair of Master Inkwells, 4½"h, heavy crystal bases, tops decorated with embossed fruit and leaves, Victorian.

PLATE 36. Pair of Copper Award Plaques: Left, inscribed "July 19th, 1922, High Jump Senior, 1st Prize"; Right, "Annual Sports Meeting, High Jump Junior, July 21, 1920".

PLATE 37.
Loving Cup,
13"h, two
handles.

PLATE 38. Souvenir
Bank from The Plains
National Bank, Lubbock,
Texas.

PLATE 39. Cowboy Hat Ash Tray, commemorative of the 1926 Dallas Cotton Palace.

PLATE 40. Match Holder in form of a cowboy boot, 4"h.

41

PLATE 41. Snuff Case, 3½″l, 3″w, copper, brass, and silver; shovel, 2″l.

PLATE 42. Gun Powder Flask, 7½″l, embossed dead game, brass fittings.

Household Copper--Kitchen & Dining Room

PLATE 43. Coffee Pot, 9"h, wooden handle and finial, hinged spout cover, embossed fleur-de-lys on spout, French or Canadian origin, lacquered.

PLATE 44. Coffee Pot, 10"h, engraved and repousse' work, lacquered.

PLATE 45. Coffee Pot, 11½"h, wooden handle, hinged brass spout cover, "Majestic" (brand) embossed on lid, lacquered.

PLATE 46. Coffee Pot, 12"h, goose-neck spout, brass finial, extended wooden handle, ca. mid-1800s, English.

PLATE 47. Coffee Pot, 10"h, wooden handle, brass finial, hinged spout cover, marked "Rome Metalware" on base, battered and mended.

PLATE 49. Perculator, 7"h, small four-cup size.

PLATE 48. Coffee Pot, 10"h, brass lid, engraved body design, originally plated.

PLATE 50. Perculator, urn style, electric, marked "Manning Quality Bowman, Meriden, Conn., patented 1910," lacquered, originally silver plated; covered Sugar and Creamer.

PLATE 51. Perculator, urn style, brass stand with burner, 14"h overall, marked "Patented March 20, 1906, and July 17, 1906," lacquered.

PLATE 52. Perculator, urn style, copper and brass, electric, marked "Universal," lacquered.

PLATE 53. Coffee Pot, 13"h, wooden handle and finial, Art Deco shape.

PLATE 54. Coal Carrier, 39"l, fancy pierced lid, English, 19th century.

PLATE 55. Coal Carrier, 33½"l, iron and wooden handle, simple punched work on lid, English, 19th century.

PLATE 56. Warming Oven, 11"h, 23½"l, hinged top, English.

50

PLATE 57. Chestnut Roaster, 23″l, iron handle, hand-punched design on lid.

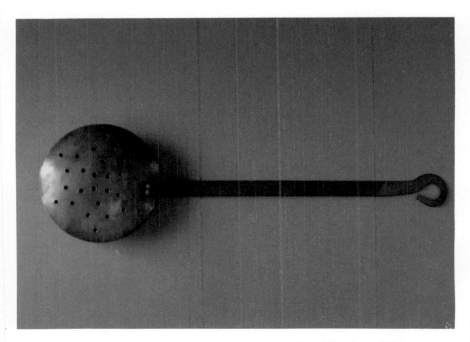

PLATE 58. Chestnut Roaster, iron handle, 16″l, pierced lid.

51

PLATE 59. Colander, 26″l, 12½″d, tin lined, pierced work forms design on base.

PLATE 60. Tinderbox Base, 3″h, 6″d.

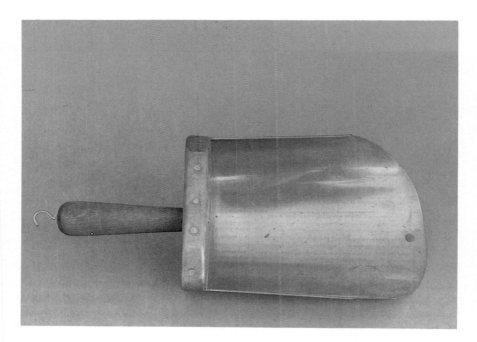

PLATE 61. Scoop, 14″l, 6″w, lacquered.

PLATE 62. Scoops made into candle sconces.

PLATE 63. Two Part Strainer or Sieve, 6″h, 12½″d, marked "Caskell & Chambers Friar Filter."

PLATE 64. Tea Caddy, 6″h, simple engraved pattern on front, badly dented.

PLATE 65. Dipper,
7½"d, 26"l, brass
handle.

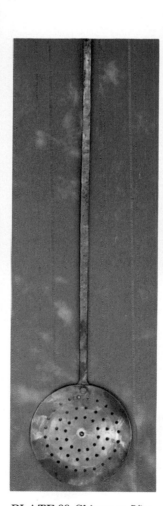

PLATE 66. Skimmer, 5"w,
22"l, pierced bowl, copper
handle.

PLATE 67. Skimmer,
7½"d, 34"l, iron
handle.

PLATE 68. Apple Butter Kettle, 13″h, 21½″d, set in three-legged iron stand, with wooden stirrer, American, mid-1800s.

PLATE 69. Candy Pan, 14″d, iron handles, lacquered.

PLATE 70. Preserving Kettle, 25″d, iron handles, soldering visible around bottom half of pan.

PLATE 71. Preserving Kettle, 26"d, iron handles.

PLATE 72. Preserving Kettle, iron bail handle, brass fittings, Argentina origin, ca. mid 1800s.

PLATE 73. Preserving Kettle, 4″h, 8″d, iron handles, ca. mid-1800s.

PLATE 74. Preserving Kettle, 8″h, 13″d, dovetailed on sides, simple soldering on base.

PLATE 75. Preserving Kettle, 18″h, 30″d, hammered body.

PLATE 76. Jelly Kettle or Pail, tin lined, 11″h, 12½″d.

PLATE 77. Preserving Kettle, 11″h, 23″d, cast iron reinforced rim.

PLATE 78. Stewing Kettle, 5½″h, 8″d.

61

PLATE 79. Stewing Kettle, 6″h, 9″d, Mid-East origin.

PLATE 80. Stewing Kettle, 8½″h, 13½″d, dovetailed construction.

PLATE 81. Stewing Kettle. 11"h, 14"d, soldered on base.

PLATE 82. Stewing Kettle, 10"h, 12"d, narrow, notched dovetailed work near base.

PLATE 83. Stewing Kettle, 9″h, 15½″d, repaired base, American, 19th century.

PLATE 84. Stewing Kettle, 8½″h, 22″d, fancy forged iron bail.

PLATE 85. Pail. 9½"h, 11"d, hammered surface, note simple brass soldered seam.

PLATE 86. Syrup Jug with funnel spout, American, New England, 19th century.

PLATE 87. Hot Liquid
Measure, 7"h.

PLATE 88. Measure, 8"h, tin
lined, marked "Kreamer."

PLATE 89. Set of Hot Liquid Measures, originally tinned hammered surfaces, lacquered: left, 5″h; center, 3½″h; right, 3″h.

PLATE 90. Measures in graduated sizes, tin lined, lacquered. Sizes: 10″h; 5½″h; 4½″h; 3¼″h; 2¾″h; 2″h. The measures are similar but not a matched set; one (5½″h) marked "Nosco."

PLATE 91: Double Measure, 8¼"l, pint (bottom) and half-pint (top) capacity.

PLATE 92. Measure, 5½"h, tin lined, brass trim.

PLATE 93. Measure, 2½"h, marked "½" (cup); Measure, 2"h, marked "⅓" (cup).

PLATE 94. Mold, 9″h, marked "Dartnall," English, 19th century.

PLATE 95. Mold, 7½"h, rose and leaf design, English.

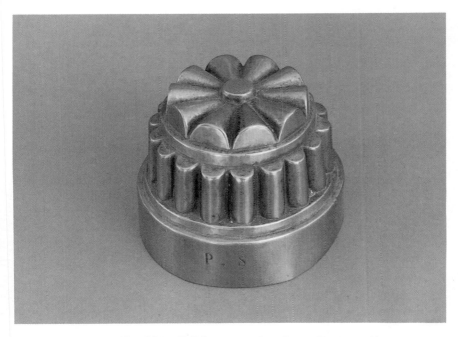

PLATE 96. Mold, 4"h, 5"d, "P.S." engraved on base, English, 19th century.

71

PLATE 97. Mold, 4″h, 4″d, marked "J.L. & Co.," English, 19th century.

PLATE 98. Mold, cross form, 6″h, English, 19th century.

PLATE 99. Mold, 5"h, 6"d, English, 19th century.

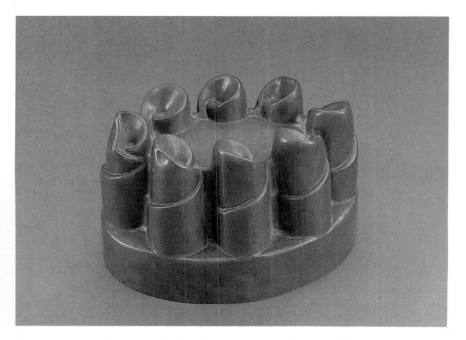

PLATE 100. Mold, 7¼"l, 5"d, oval shape, English, 19th century.

PLATE 101. Tube Mold, 4″h, 8″d, German, late 19th century.

PLATE 102. Mold with crockery insert and wire bail handle, early 20th century.

PLATE 103. Tube Mold, 2½"h, 5½"d, 20th century.

PLATE 104. Ring Mold, 6"d, English, 19th century.

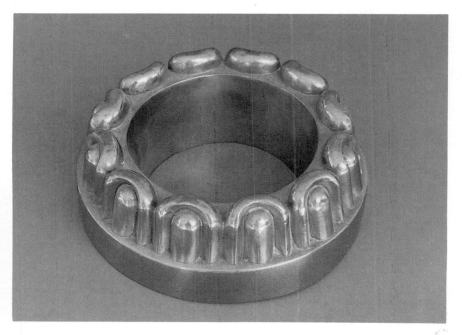

PLATE 105. Ring Mold, 2½"h, 7"d, marked "273," Art Deco design, English.

PLATE 106. Oval Mold, tin base and collar, grape pattern, English, 19th century.

PLATE 107. Mold, tin base (collar missing), crown pattern, English, 19th century.

PLATE 108. Mold, 10″l, fruit basket design, 20th century.

PLATE 109. Tube Mold, 5"h, six-sided, floral pattern on top, English, 19th century.

PLATE 110. Tube Mold, 5"h, round base with top shaped into six sides, English, 19th century.

PLATE 111. Tube Mold, 10½″d, German, early 20th century.

PLATE 112. Mold, 3″h, 4¼″d, German, hammered surface, early 20th century.

PLATE 113. Mold, 11½″d, simple design, early 20th century.

PLATE 114. Tube Mold, 4"h, 7"d, 20th century.

PLATE 115. Tube Mold, 11"d, early 20th century.

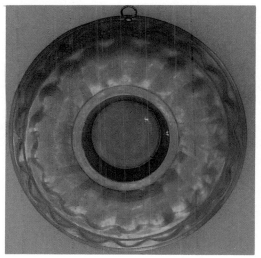

PLATE 116. Mold, 5¼"h, 9"d, cross design, wire bail handle, German, early 20th century.

PLATE 117. Milk Pitcher, 7½"h, 6½"d, hinged lid, Russian, 19th century.

PLATE 118. Pitcher, 8½"h, Art Nouveau style, English.

PLATE 119. Syrup Pitcher, 6"h, dovetailed construction, 19th century.

PLATE 120. Water Can, 9"h, tin top, ca. mid-1800s.

PLATE 121. Pitcher, 9"h, tin lined, ca. mid-1800s.

PLATE 122. Pitcher, 15"h, engraved design of Lion and Unicorn with old French saying "Honi Soit Qui May Y Pense" (evil be to him who thinks evil), marked "England" and "Celebrate."

PLATE 123. Pitcher or Jug, 7″h, hammered body, tin lined.

PLATE 124. Set of Colonial Mexican Pitchers, 8″, 7″, 6″ in height, hammered bodies, dovetailed, tin lined.

83

PLATE 125. Stock Pot, 19½″h, tin lid.

PLATE 126. Stock Pot, 13″h, 10″d, brass handles, stamped "Clarendon, Hammersmith," English, 19th century.

PLATE 127. Stock Pot, 20"h, wooden handles, brass and copper spigot, lacquered.

PLATE 128. Stock Pot, 20½"h, 19"d, brass handles, spigot missing, English, 19th century.

PLATE 129. Steamer, 24″l, 15″w, 10″h, iron handle, English, 19th century.

PLATE 130. Vegetable Steamer, 8½″h, 14″d, brass handle, English.

PLATE 131. Steamer, 5"h, 15"l, with drainer (not shown).

PLATE 132. Fish Kettle (lid missing), 14"l, 9"d, tin lined, brass handles, English, mid-1800s.

PLATE 133. Clam Steamer, 11″h, 8″d, marked "Fried, Loblich, Wein," Austrian.

PLATE 134. Covered Sauce Pan, 10½"h, 10"d. Lid is marked "F.A. Walker
& Co., Cornhill, Boston, Mass," hammered surface, iron handles, lacquered, ca.
early 20th century.

PLATE 135. Sauce Pan, 9"h, 6"d, iron handle, lacquered, marked "P. Palmier,
Ortega 34, Mexico," 20th century.

PLATE 136. Sauce Pan, 5¼″h, 9″d, dovetailed on base, iron handle, 19th century.

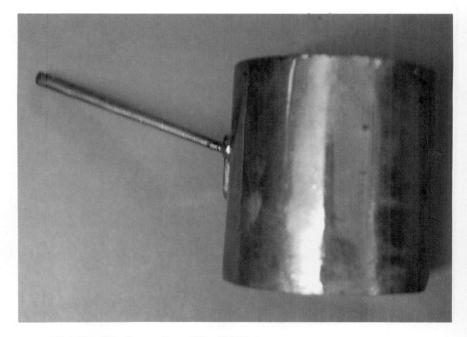

PLATE 137. Sauce Pan, 7″h, 6½″d, brass handle, 19th century.

PLATE 138. Covered Sauce Pan, 11"h, 14"d, iron handles, dovetailed, 19th century.

PLATE 139. Sauce Pan, 5"h, 9½"d, brass handle.

PLATE 140. Sauce Pan, 3″h, 12″d, dovetailed, New England, 19th century.

PLATE 141. Set of Sauce Pans: 7¼″d; 5½″d; 5″d; 4½″d;
3½″d; brass handles, French, ca. mid-20th century.

PLATE 142. Sauce Pan, 5″h, 8″d, dovetailed, iron handle, marked "C.C. Mutual, Catherine, N.Y."

PLATE 143. Skillet, 2½″h, 9½″d, dovetailed, hammered surface, iron handle, lacquered, 19th century.

PLATE 144. Skillet, 2″h, 10¼″d, hammered body, iron handle.

PLATE 145. Ebleskiver Pan, 3″h, 11″d.

PLATE 146. Ebleskiver
Pan, 3"h, 11"d, ring hook.

PLATE 147. Escargot
Pan, 2½"h, 18"d.

95

PLATE 148. Baking Pan, 5½″h, 8″d, tin lined, brass handles.

PLATE 149. Roasting Pan, 27″l, 20″w, 7″h, diamond-shaped, English, 19th century.

PLATE 150. Baking Pan, 2½"h, 11½"d, ring hook.

PLATE 151. Two-Part
Baking Pan or Cooker,
5"h, 8"d, tin lined. Top
could serve as lid,
separate pan, or plate.

PLATE 152. Baking Pan, 2½″h, 11½″d, iron ring hook, hammered surface, originally tinned.

PLATE 153. Baking Pans: Left, 3″h, 4¼″d; Right, 3½″h, 5¼″d. Left is marked "G. Fontana, London."

PLATE 154. Set of Baking Pans: 4½″ x 11½″; 3¾″ x 10″; 3½″ x 8″.

PLATE 155,156,157. Early 20th century American Tea Kettles, originally nickel plated. These kettles were made by several manufacturers in various sizes. Plate 155, 8″h, 13″d, marked "Revere Ware." Plate 156: Left, 6″h, 8″d; Right, 7″h, 10″d, porcelain finials. Plate 157, 5½″h, 6½″d, lacks lid, unusual small size.

PLATE 158. Kettle, 7″h, tin handle, figural bird finial on spout cover, Art Deco embossed design on side.

PLATE 159. Kettle, 10″d, wooden and iron handle, wooden finial, marked "Majestic."

PLATE 160. Kettle marked "Pilgrim Ware, 2 Rivers, Wis.. Pat. Applied for."

PLATE 161. Kettle, wooden and copper handle, copper knob finial.

PLATE 162. Kettle, 9"h, 6½"d, fancy wooden and brass handle, brass finial, marked "China."

PLATE 163. Kettle, 10½″h, 10″d, lacquered, English, 19th century.

PLATE 164. Kettle, 9½"d, 9½"h, wooden finial, English.

PLATE 165. 11"h, 8"d, goose-neck spout, brass finial, brass and copper handle, 19th century.

PLATE 166. Kettle, 13″h, 6″d, goose-neck spout, brass and copper handle, brass spout, acorn finial, Victorian.

PLATE 167. Kettle, 11″h, 9″d, brass handle, acorn finial, lacquered.

PLATE 168. Kettle, 11½″h, 7½″d, riveted wire bail handle, European.

PLATE 169. Kettle, 8½″h, 6″d, brass handle and finial, hinged lid, European, 19th century.

PLATE 170. Kettle, 8½"h, 6"d, broad, flat handle, brass finial, English, 19th century.

PLATE 171. Kettle, 11½"h, 8"d, oval shape, goose-neck spout, 19th century.

PLATE 172. Kettle, 10½"h, 8"d, American (Frankfurt, Indiana), note wide dovetailed work around top.

PLATE 173. Kettle, 12"h, 6"d, goose-neck spout, brass and copper handle, brass acorn finial, marked "SE," 19th century.

PLATE 174. Kettle, 11½"h, 9"d, oval shaped, goose-neck spout, 19th century.

PLATE 175. Kettle, with stand and burner, initialed "C.D." on base, Victorian.

PLATE 176. Kettle with burner and pewter stand, 14½"h overall, bamboo shaped spout and legs. Victorian.

PLATE 177. Kettle, 7½"h, brass stand, burner with cover, Victorian.

PLATE 178. Kettle, 6½"h, brass and copper; brass stand, burner missing, ca. mid-20th century.

PLATE 179. Tea Urn, 18″h, brass trim; Tray, 24″l, 16″w, English, 19th century.

PLATE 180. Wine Cooler, 7″h, lacquered, originally silver plated, marked "Rep. Sheffield, Industrial Argentina," engraved Art Nouveau design on front.

111

PLATE 181. Egg Warmer, 10″h, pewter trim. English, Victorian.

PLATE 182. Plate Warmer, 11″d. Hot water was poured into extended opening on side.

PLATE 183. Plate Warmer with lid, 9″l, oval shaped, brass handles.

PLATE 184. Trivet, 10″d (expands to 13″d), engraved and pierced work, marked "Royal Rochester, Sheffield."

PLATE 185. Chafing Dish, 12″h overall, pan, 9″d, copper and brass, wooden handles, ca. mid-20th century.

PLATE 186. Warming Tray, 6"h, 17"l, brass handles and legs, marked with an anchor design.

PLATE 187. Warming Tray, 33"l, 9½"w, marked "Alex Boyd & Son, Summers, 105 New Bond St., London."

PLATE 188. Centerpiece Bowl, 10″h, 11″d, lion head with ring handles, originally plated.

PLATE 189. Centerpiece, 12″d, pot metal Art Deco figure and birds, originally plated.

PLATE 190. Divided dish, 11¼″d, white metal border, originally plated, ca. mid-20th century.

PLATE 191. Tray, 20½″l, 9¼″w, embossed berry and leaf pattern.

117

PLATE 192. Tray, 16″l, wood and copper, engraved work, English.

PLATE 193. Dinner Gong, 24″d, marked "Burmese." (Mounted on wooden brush holder)

118

PLATE 194. Bowl, 2"h, 5"d, blue enameled interior, marked "China."

PLATE 195. Compote, 4"h, 9"w, pierced work on outer border, originally plated.

PLATE 196. Tray, 15½"d, Art Nouveau design around border, originally plated.

119

PLATE 197. Candleholder, 9¼″h, brass repoussé work, originally plated.

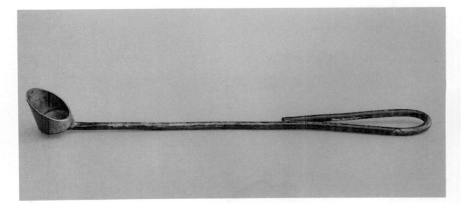

PLATE 198. Candle Snuffer, 12″l, wrought iron handle.

Household Copper--Bed & Bath

PLATE 199. Chamberstick, 4½"h, 5½"d, copper and brass, originally plated.

PLATE 200. Chamberstick, 4½"h, 5½"d, brass handle, 19th century.

PLATE 201. Warming Pan, 41″l, 13″d, brass top.

PLATE 202. Warming Pan, 42″l, 11″d, lacquered.

PLATE 203. Warming Pan, 40″l, 13″d.

PLATE 204. Warming Pan, 46″l, 13″d, lacquered.

PLATE 205. Bed Warmer, 12″d, brass stopper with ring handle, lacquered.

PLATE 206. Bed Warmer, 18″l, oval shaped, brass fittings, English, ca. mid-19th century.

PLATE 207. Bed Warmer, 8½"d, impressed marks around top, "Wafax, Fill Completely, Reg. Design."

PLATE 208. Bed Warmer, 9"d, brass stopper, marked "Fill to Rim."

PLATE 209. Bed Warmer, 7"d, marked "Wendy, Fill to Rim."

PLATE 210. Lavabo and Cistern, brass handles and spigots, lacquered, 19th century.

PLATE 211. Cistern, 22″h, brass acorn finial on lid, brass spigot, English, 19th century.

PLATE 212. Cistern, 15″h, 19″l, oval shaped, brass finial and spigot.

PLATE 213. Shower Head, 8″d, pierced work around rim.

PLATE 214. Electric Heater, 11½″d, marked "The Ac. Gilbert Co., New Haven, Conn., U.S.A."

PLATE 215. Geyser, 29"h, marked "New Geysers, Ltd., London, Barralet Patents."

PLATE 216. Geyser, 29"h, marked "Edwarts, Victor-Geyser, London."

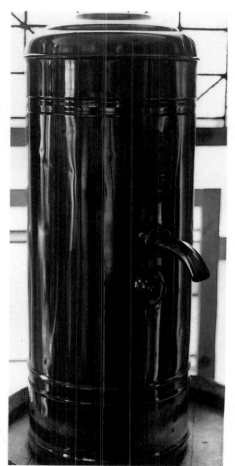

PLATE 217. Child's Tub, tin base, lacquered, English.

Household Copper--Laundry & Cleaning

PLATE 218. Washing Machine, electric with reversible wringer, marked "Judd Laundry Machine Co., Chicago, Ill., Oct. 12, 1909."

PLATE 219. Wash Tub, marked "Perco Washer, Mystic Washer Co., Houston, Texas, Patent Pending."

PLATE 220. Wash Tub and Lid, 18"h, marked "Mystic Washer, Mfd. by The Kettler Mfg. Co., Houston, Texas, Patent. Pending," tin lined.

PLATE 221. Tub, 13"h, 12"d, tin lined.

PLATE 222. Tub, 12"h, 14"d, crude soldering visible around top.

PLATE 223. Tub, 12"h, tin lined.

PLATE 224. Wash Boiler, marked "Atlantic 11 Gallon."

PLATE 225. Wash Boiler, 13″h, 24″l, marked "Sullivan Geiger Co., Indianapolis."

PLATE 226. Wash Boiler, 14″h, 27″l, marked "Canco."

PLATE 227. Wash Boiler, 13″h, 23″l, marked "Nesco."

135

PLATE 228. Wash Boiler, 13½″h, 26½″l, marked "Rochester," tin lid.

PLATE 229. Wash Boiler, 13½″h, 28½″l, copper lid.

PLATE 230. Wash Boiler, 13"h, 25"l, unmarked.

PLATE 231. Tub, 18"h, 27"d, lacquered.

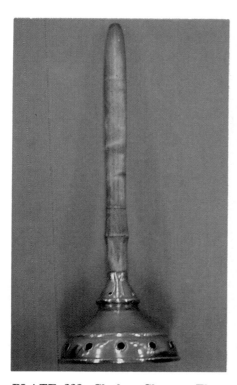

PLATE 232. Clothes Cleaner. These were used to force hot water through the clothes.

PLATE 234. Rug Beater, 30"l, wooden handle.

PLATE 233. Dust Pan, 9"l, 8¼"w, embossed designs in Art Nouveau style.

138

Commercial Copper

PLATE 235. Barber's Hot Water Container, 14″h, 11″d, contains tray for sterilizing instruments, lacquered.

PLATE 236. Double Coffee Urn, 35½"h, 25"w, three spigots, lacquered, marked "Smith St. John, Kansas City."

PLATE 237. Cappucino Machine (Italian Coffee drink), 43"h, 31"w, brass trim, marked "Brevetti Gaggia," Italian.

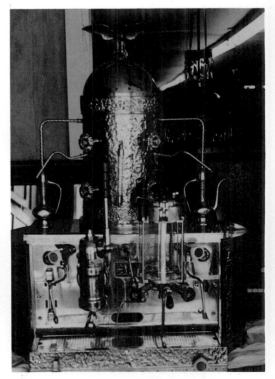

PLATE 238. Coffee Urn with crockery servers, 14½″h overall, unmarked, probably German.

PLATE 239. Coffee Grinder and Dispenser, brass lid, marked "Neckzugel, Wein," Austrian.

PLATE 240. Commercial Hot Water (or liquid) Container, 13"h, 25"l, tin top, hinged, porcelain finial, brass spigot, lacquered.

PLATE 241. Hot Water Container, rectangular shape, iron handles, lacquered.

PLATE 242. Hot Water Container, 12″h, 11″d, brass spigot, iron handles, lacquered.

PLATE 243. Still, 24″h, 18½″d, brass spigot, lacquered.

143

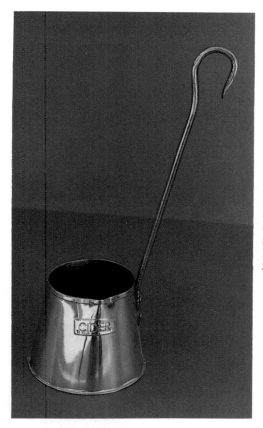

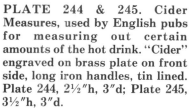

PLATE 244 & 245. Cider Measures, used by English pubs for measuring out certain amounts of the hot drink. "Cider" engraved on brass plate on front side, long iron handles, tin lined. Plate 244, 2½"h, 3"d; Plate 245, 3½"h, 3"d.

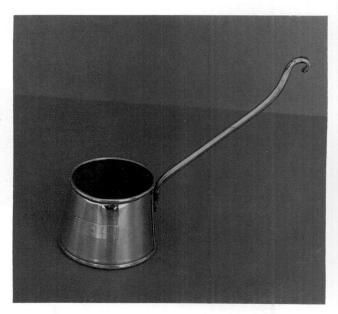

PLATE 246. Ale Warmer, 7½″h, 8″l, English, 19th century. The extended base was heated before serving to keep the drink warm.

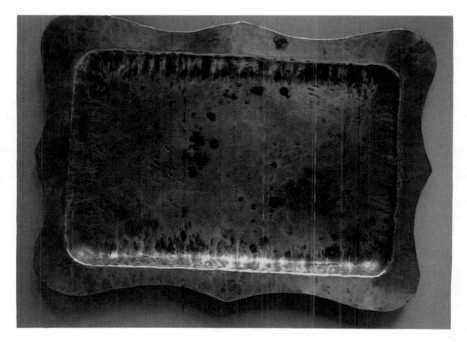

PLATE 247. Tip Tray, 7¼″l, 5″w, hammered surface.

PLATE 248. Restaurant Tray, 12″d, "Jamie's" embossed in brass on border, circa mid-20th century.

PLATE 249. Cigarette Receptacle, 8″h, 12½″d, brass handles. The box holds fine grained sand.

PLATE 250. Merchant's Scale with copper shelf, marked "Doyle & Son, Borough, London, SE 1."

PLATE 251. Merchant's Scale with copper pan, iron weights, marked capacity "To Weigh 14 lbs," 8″h overall.

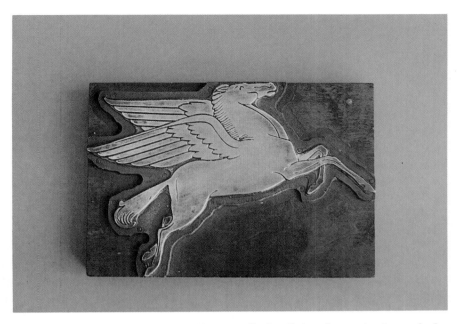

PLATE 252. Printer's Plate, "Pegasus," the flying horse trade-mark for Magnolia Oil Co.

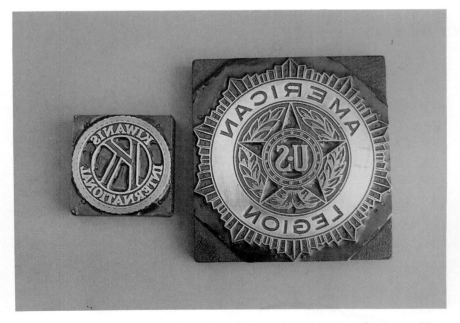

PLATE 253. Printer's Plate, Kiwanis emblem on left: American Legion emblem on right.

PLATE 254. Printer's Plate, "St. Louis Flower Show, 1912, Coliseum."

PLATE 255. Printer's Plate, "Hart, Schafner & Marx."

PLATE 256. Printer's Plate, advertisement for Waterspar Paint.

149

PLATE 257. Film Developing Tank, brass fittings.

PLATE 258. Sterilizer, 18″l, insert (not shown), brass handles.

PLATE 259. Mold for Rubber Boot, 10½″h.

PLATE 260. Mold for Rubber Boot, 14½″h.

PLATE 261, 262, 263. Molds for Leather Cowboy Boots: Plate 261, 10½″h; Plate 262, 13″h; Plate 263, 7½″h (child's size).

152

PLATE 264. Fire Extinguisher, 24"h (made into a lamp), marked "Quick Aid, Detroit Corp. Detroit, Michigan."

PLATE 265. Fire Extinguisher, hand pumped for pressure, 18½"h, marked "Loestrand, Rockville, Md."

PLATE 266. Railroad Lantern, 21″h, marked "BR (W)."

PLATE 267. Lantern, 9″h, marked "S.N.L.W., Ltd.," English.

PLATE 268. Railroad Lantern, 12″h, marked "BR."

PLATE 269. Railroad Torch, 6″h.
The spout held a wick.

PLATE 270. Glue Pot, 6½″h,
brass handles and burner attach-
ment, electric.

PLATE 271. Industrial Container (unknown function), 12½"h, lacquered.

PLATE 272. Yacht Ventilator, brass base.

PLATE 273. Oil Can, 4″h, 3½″w, French.

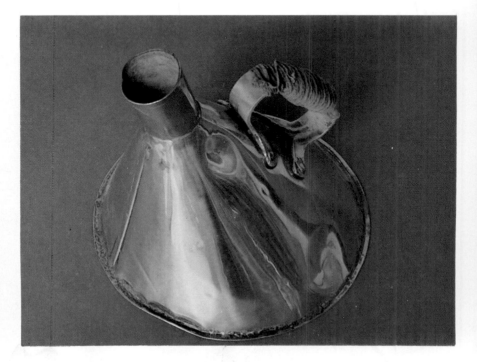

PLATE 274. Oil Can, 9″h, 9″d.

PLATE 275. Industrial Funnels, 24"h, 12½"d when joined.

PLATE 276. Funnel, 10½"h, 12"d.

159

PLATE 277. Funnel, 8½″h, 6″d, brass plunger.

PLATE 278. Funnel, 5½″h.

160

PLATE 279. Funnel, 12½″h.

PLATE 280. Flash Light, 8″l, marked "Ray-o-Vac."

PLATE 281. Cow Bell, 7″h.

PLATE 282. Soldering Iron, 13″l, copper, iron, and wood.

PLATE 283. Pump, 32″l, marked "Cathol, Mass."

PLATE 284. Spray Pump, 14″l, copper and brass.

PLATE 285. Spray Pump, 19½″l, amber glass base.

Current Reproductions

PLATE 286. Ebleskiver Pan, 8½"d, wooden handle, lacquered.

PLATE 287. Ebleskiver Pan, 7"d, iron handle.

PLATE 288. Chocolate Mold, 13"d, brass ring hook.

PLATE 289. Set of Cider Measures, brass name plates, iron handles, unlined, lacquered.

PLATE 290. Chocolate Mold, 15″l, 3¼″w, lion figures, copper plated.

PLATE 291. Ash Tray, 3½″d, brass paw feet, faked "verdigris."

PLATE 292. Cuspidor, 10″h, brass interior, marked "All Famous Havana 5¢ cigars," lacquered.

PLATE 293. Preserving Kettle, 11″h, 19″d.

PLATE 294. Umbrella Stand, 21″h, brass handles and base, lacquered.

PLATE 295. Coal Scuttle, 10″h, 16″d, "dimpled" surface.

PLATE 296. Dust Pan to match Scuttle, 13½″l, 9″w.

PLATE 297. Coat Scuttle, 16″h, hammered body, brass pedestal base, "delft-style" ceramic handle, lacquered.

168

PLATE 298. Weather Vane, faked "verdigris" on compass points.

PLATE 299. Coal Scuttle, 17"h, ceramic "delft-style" handle, applied lion heads on each side.

169

PLATE 300. Lantern, marked "Anchor," brass fittings.

PLATE 301. Ship's Light, electric, 9¼"h, marked "Port" and "Hop Lee & Co., Hong Kong, No. 4751" (on brass plates on top)

170

Bibliography

Atterbury, Paul (ed.). *An Encyclopedia of the Decorative Arts*. London: Octopus Books, Ltd., 1979.

Baker, Stanley L. *Railroad Collectibles*, 2nd ed. Paducah, Kentucky: Collector Books, 1981.

Curtis, Anthony (comp.). *The Lyle Antiques & Their Values, Metalwork Identification & Price Guide*. Voor Hoede Publications B.V., 1982.

Dreppard, Carl. "Paul Revere, Brass and Coppersmith," in Albert Revi (ed.), *Collectible Iron, Tin, Copper & Brass*. Everybody's Press, 1974.

Franklin, Linda Campbell. *300 Years of Kitchen Collectibles*. Florence, Alabama: Books Americana, 1981.

Gaston, Mary Frank. *Antique Brass*. Paducah, Kentucky: Collector Books, 1985.

Gentle, Rupert & Rachael Feild. *English Domestic Brass*. New York: E.P. Dutton and Co., Inc., 1975.

Haedeke, Hanns-Ulrich (translated by Vivienne Menkes). *Metalwork*. New York: Universe Books, 1969.

How Things Work. Vol. III. Geneva: Bibliographisches Institute and Simon and Schuster Inc., American Edition, n.d.

Kauffman, Henry J. "Early American Brass and Copper and its makers," pp. 104-107 in Albert Revi (ed.), *Collectible Iron, Tin, Copper & Brass*. Everybodys Press, 1974.

_____. "Collecting American Copper," pp. 56-58 in *Antique Trader Weekly*, December 9, 1981.

Ketchum, William C. *American Antiques*. New York: Rutledge Books, Inc., 1980.

_____. *Western Memorabilia*. New Jersey: Hammond, 1980.

Moore, N.H. *Old Pewter, Brass, Copper & Sheffield Plate*. Garden City, New York: Garden City Publishing Company, Inc., (1905) 1933.

Perry, Evan. *Collecting Antique Metalware*. Garden City, New York: Doubleday & Company, Inc., 1974.

Sears, Roebuck Catalog 1908. Chicago Illinois: The Gun Digest Company, 1969.

Thuro, Catherine. *Primitives & Folk Art*. Paducah, Kentucky: Collector Books, 1979.

Wills, Geoffrey. *Collecting Copper & Brass*. England: Arco Publications, 1962.

_____. *The Book of Copper and Brass*. Feltham, England: The Hamylin Publishing Group Limited for Country Life Books, 1968.

Other Books by Mary Frank Gaston

The Collector's Encyclopedia of Limoges Porcelain $19.95
The Collector's Encyclopedia of R.S. Prussia $24.95
Blue Willow, An Illustrated Value Guide $9.95
The Collector's Encyclopedia of Flow Blue China $19.95
Haviland Collectables and Objects of Art $19.95
American Belleek $19.95
Brass, Identification and Values $9.95

These titles may be ordered from the author or the publisher Include $1.00 each for postage and handling.

Mary Frank Gaston   Collector Books
P.O. Box 342   P.O. Box 3009
Bryan, Texas 77806   **171**   Paducah, KY 42001

Object Index to Photograph Numbers

Price Guide

PLATE 1	$95.00	PLATE 52	$495.00
PLATE 2	$42.00	PLATE 53	$125.00
PLATE 3	$30.00	PLATE 54	$235.00
PLATE 4	$125.00	PLATE 55	$175.00
PLATE 5	$175.00	PLATE 56	$375.00
PLATE 6	$450.00	PLATE 57	$85.00
PLATE 7	$2200.00	PLATE 58	$65.00
PLATE 8	$70.00	PLATE 59	$225.00
PLATE 9	$150.00	PLATE 60	$30.00
PLATE 10	$600.00	PLATE 61	$125.00
PLATE 11	$225.00	PLATE 62	(left) $65.00
PLATE 12	$350.00		(right) $47.50
PLATE 13	$175.00	PLATE 63	$120.00
PLATE 14	$195.00	PLATE 64	$55.00
PLATE 15	$125.00	PLATE 65	$125.00
PLATE 16	$90.00	PLATE 66	$60.00
PLATE 17	$685.00	PLATE 67	$40.00
PLATE 18	$1700.00	PLATE 68 Kettle & Stand	$425.00
PLATE 19	$250.00		Stirrer $35.00
PLATE 20	$375.00	PLATE 69	$75.00
PLATE 21	$75.00	PLATE 70	$300.00
PLATE 22	$90.00	PLATE 71	$285.00
PLATE 23	$65.00	PLATE 72	$100.00
PLATE 24	$225.00	PLATE 73	$125.00
PLATE 25	$37.50	PLATE 74	$110.00
PLATE 26	$15.00	PLATE 75	$300.00
PLATE 27	$80.00	PLATE 76	$265.00
PLATE 28	pair $295.00	PLATE 77	$225.00
PLATE 29	pair $225.00	PLATE 78	$125.00
PLATE 30	$265.00	PLATE 79	$47.50
PLATE 31	set $45.00	PLATE 80	$95.00
PLATE 32	$40.00	PLATE 81	$95.00
PLATE 33	each $15.00	PLATE 82	$95.00
PLATE 34	$75.00	PLATE 83	$215.00
PLATE 35	each $175.00	PLATE 84	$175.00
PLATE 36	each $13.00	PLATE 85	$200.00
PLATE 37	$125.00	PLATE 86	$150.00
PLATE 38	$17.50	PLATE 87	$95.00
PLATE 39	$5.00	PLATE 88	$60.00
PLATE 40	$17.50	PLATE 89	$125.00; $100.00; $75.00
PLATE 41	$285.00	PLATE 90	(back row) $100.00;
PLATE 42	$182.00		$125.00; $75.00; (front
PLATE 43	$125.00		row) $60.00; $50.00; $40.00
PLATE 44	$95.00	PLATE 91	$125.00
PLATE 45	$55.00	PLATE 92	$100.00
PLATE 46	$290.00	PLATE 93	(left) $20.00
PLATE 47	$25.00		(right) $15.00
PLATE 48	$45.00	PLATE 94	$350.00
PLATE 49	$38.50	PLATE 95	$200.00
PLATE 50	set $595.00	PLATE 96	$150.00
PLATE 51	$165.00	PLATE 97	$150.00

PLATE 98	$200.00	PLATE 152	$155.00
PLATE 99	$275.00	PLATE 153	(left) $45.00
PLATE 100	$350.00		(right) $55.00
PLATE 101	$125.00	PLATE 154(large to small) $100.00;
PLATE 102	$175.00	$85.00; $70.00	
PLATE 103	$75.00	PLATE 155	$40.00
PLATE 104	$125.00	PLATE 156	(left) $37.50
PLATE 105	$175.00		(right) $45.00
PLATE 106	$175.00	PLATE 157	$69.00
PLATE 107	$150.00	PLATE 158	$65.00
PLATE 108	$90.00	PLATE 159	$49.50
PLATE 109	$200.00	PLATE 160	$125.00
PLATE 110	$225.00	PLATE 161	$40.00
PLATE 111	$95.00	PLATE 162	$135.00
PLATE 112	$28.00	PLATE 163	$155.00
PLATE 113	$90.00	PLATE 164	$115.00
PLATE 114	$100.00	PLATE 165	$105.00
PLATE 115	$95.00	PLATE 166	$95.00
PLATE 116	$125.00	PLATE 167	$150.00
PLATE 117	$150.00	PLATE 168	$65.00
PLATE 118	$66.00	PLATE 169	$120.00
PLATE 119	$45.00	PLATE 170	$80.00
PLATE 120	$195.00	PLATE 171	$225.00
PLATE 121	$75.00	PLATE 172	$220.00
PLATE 122	$150.00	PLATE 173	$195.00
PLATE 123	$150.00	PLATE 174	$225.00
PLATE 124	$75.00;$60.00;$45.00	PLATE 175	$195.00
PLATE 125	$225.00	PLATE 176	$180.00
PLATE 126	$375.00	PLATE 177	$125.00
PLATE 127	$175.00	PLATE 178	$115.00
PLATE 128	$850.00	PLATE 179	set $2500.00
PLATE 129	$490.00	PLATE 180	$110.00
PLATE 130	$125.00	PLATE 181	$145.00
PLATE 131	$165.00	PLATE 182	$175.00
PLATE 132	$700.00	PLATE 183	$250.00
PLATE 133	$135.00	PLATE 184	$50.00
PLATE 134	$350.00	PLATE 185	$128.00
PLATE 135	$175.00	PLATE 186	$325.00
PLATE 136	$65.00	PLATE 187	$225.00
PLATE 137	$200.00	PLATE 188	$50.00
PLATE 138	$445.00	PLATE 189	$285.00
PLATE 139	$185.00	PLATE 190	$55.00
PLATE 140	$225.00	PLATE 191	$95.00
PLATE 141	set $225.00	PLATE 192	$42.50
PLATE 142	$105.00	PLATE 193	$175.00
PLATE 143	$275.00	PLATE 194	$25.00
PLATE 144	$175.00	PLATE 195	$27.50
PLATE 145	$185.00	PLATE 196	$39.50
PLATE 146	$195.00	PLATE 197	$80.00
PLATE 147	$179.00	PLATE 198	$30.00
PLATE 148	$125.00	PLATE 199	$70.00
PLATE 149	$360.00	PLATE 200	$45.00
PLATE 150	$145.00	PLATE 201	$250.00
PLATE 151	$65.00	PLATE 202	$185.00

PLATE 203..............$165.00	PLATE 256..............$20.00
PLATE 204..............$185.00	PLATE 257..............$25.00
PLATE 205..............$125.00	PLATE 258..............$225.00
PLATE 206..............$550.00	PLATE 259..............$175.00
PLATE 207...............$40.00	PLATE 260..............$135.00
PLATE 208...............$70.00	PLATE 261..............$145.00
PLATE 209...............$45.00	PLATE 262..............$125.00
PLATE 210.............$2750.00	PLATE 263..............$115.00
PLATE 211..............$690.00	PLATE 264..............$180.00
PLATE 212..............$175.00	PLATE 265...............$50.00
PLATE 213...............$65.00	PLATE 266..............$130.00
PLATE 214...............$47.50	PLATE 267...............$75.00
PLATE 215..............$175.00	PLATE 268...............$60.00
PLATE 216..............$140.00	PLATE 269...............$50.00
PLATE 217..............$300.00	PLATE 270..............$125.00
PLATE 218.............$1295.00	PLATE 271...............$26.00
PLATE 219..............$125.00	PLATE 272...............$60.00
PLATE 220...............$95.00	PLATE 273...............$45.00
PLATE 221...............$60.00	PLATE 274...............$65.00
PLATE 222...............$75.00	PLATE 275..............$225.00
PLATE 223...............$95.00	PLATE 276...............$45.00
PLATE 224..............$125.00	PLATE 277...............$40.00
PLATE 225...............$85.00	PLATE 278...............$22.00
PLATE 226...............$62.50	PLATE 279..............$125.00
PLATE 227..............$110.00	PLATE 280...............$25.00
PLATE 228...............$77.50	PLATE 281...............$15.00
PLATE 229..............$110.00	PLATE 282...............$14.00
PLATE 230..............$100.00	PLATE 283..............$175.00
PLATE 231..............$195.00	PLATE 284...............$32.50
PLATE 232...............$25.00	PLATE 285...............$80.00
PLATE 233...............$35.00	
PLATE 234...............$30.00	
PLATE 235..............$275.00	
PLATE 236..............$895.00	
PLATE 237..............$600.00	
PLATE 238..............$185.00	
PLATE 239..............$575.00	
PLATE 240..............$145.00	
PLATE 241..............$175.00	
PLATE 242..............$105.00	
PLATE 243..............$390.00	
PLATE 244...............$90.00	
PLATE 245...............$90.00	
PLATE 246..............$150.00	
PLATE 247...............$10.00	
PLATE 248...............$20.00	
PLATE 249..............$210.00	
PLATE 250..............$150.00	
PLATE 251..............$175.00	
PLATE 252...............$29.00	
PLATE 253..........(left) $10.00	
..........(right) $18.00	
PLATE 254...............$25.00	
PLATE 255...............$25.00	

Two Important Tools For The
Astute Antique Dealer, Collector and Investor

Schroeder's Antiques Price Guide

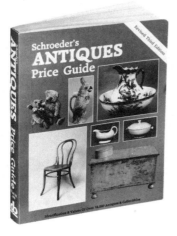

The very best low cost investment that you can make if you are really serious about antiques and collectibles is a good identification and price guide. We publish and highly recommend **Schroeder's Antiques Price Guide.** Our editors and writers are very careful to seek out and report accurate values each year. We do not simply change the values of the items each year but start anew to bring you an entirely new edition. If there are repeats, they are by chance and not by choice. Each huge edition (it weighs 3 pounds!) has over 56,000 descriptions and current values on 608 - 8½x11 pages. There are hundreds and hundreds of categories and even more illustrations. Each topic is introduced by an interesting discussion that is an education in itself. Again, no dealer, collector or investor can afford not to own this book. It is available from your favorite bookseller or antiques dealer at the low price of $9.95. If you are unable to find this price guide in your area, it's available from Collector Books, P. O. Box 3009, Paducah, KY 42001 at $9.95 plus $1.00 for postage and handling.

Flea Market Trader

Bargains are pretty hard to come by these days -- especially in the field of antiques and collectibles, and everyone knows that the most promising sources for those seldom-found under-priced treasures are flea markets. To help you recognize a bargain when you find it, you'll want a copy of the *Flea Market Trader* -- the only price guide on the market that deals exclusively with all types of merchandise you'll be likely to encounter in the marketplace. It contains not only reliable pricing information, but the *Flea Market Trader* will be the first to tune you in to the market's newest collectible interests -- you'll be able to buy before the market becomes established, before prices have a chance to escalate! You'll not only have the satisfaction of being first in the know, but you'll see your investments appreciate dramatically. You'll love the format -- its handy 5½" x 8½" size will tuck

easily into pocket or purse. Its common sense organization along with a detailed index makes finding your subject a breeze. There's tons of information and hundreds of photos to aid in identification. It's written with first-hand insight and an understanding of market activities. It's reliable, informative, comprehensive; it's a bargain! From Collector Books, P. O. Box 3009, Paducah, KY 42001 at $8.95 plus $1.00 for postage and handling.